Peter T. Forsyth

Sacramentalism the True Remedy for Sacerdotalism

Peter T. Forsyth

Sacramentalism the True Remedy for Sacerdotalism

ISBN/EAN: 9783337426958

Printed in Europe, USA, Canada, Australia, Japan

Cover: Foto ©berggeist007 / pixelio.de

More available books at **www.hansebooks.com**

THE

EXPOSITOR.

EDITED BY THE REV.

W. ROBERTSON NICOLL, M.A., LL.D.

FIFTH SERIES.

Volume VIII.

London:

HODDER AND STOUGHTON,

27, PATERNOSTER ROW.

———

MDCCCXCVIII.

SACRAMENTALISM THE TRUE REMEDY FOR SACERDOTALISM.

I.

IF it were asked how non-sacerdotalists regard the Communion rite, the first answer of many, if not most, would tend to replace the word Communion altogether by the term Commemoration, just as for the word Baptism many would substitute a term like Dedication. The result in each case is to avoid magical interpretation by emptying the rite of all mystic value; and the protest against superstition is effected by parting with all spiritual intimacy and profundity not realisable by the plain good man in the street. To some, indeed, a Sacrament is no more than an object lesson or spectacle, exhibiting certain truths in a condensed form, and clothing them with more or less impressiveness—and mostly less.

Let us come to closer quarters with our facts and truths, and assay what gold there is in this white stone ruddy-veined which we have inherited with our spiritual estate.

Let it be first observed that the Communion is an *act*. It is not simply a feeling nor a contemplation. So far it may be described as an *opus operatum*. "Do this," is the word, not, "consider this." The Saviour in that hour certainly did not think of Himself æsthetically, as an object of contemplation. Nor is it simply "remember Me." The reminiscence is subordinated to the act enjoined. It is more than a reminiscence; it is at least a reminiscent act. The very variations in the form of instituting words only direct attention on the centre of the occasion as an act. Something is done. It is the worship of bowed ·wills even more than of changed hearts. And its expression is less the streaming eye of emotion than the bent head of obedience and obeisance.

Moreover, it is an act *of the Church* more than of the individual. It was not to a group of individuals that the command was given, but to a body already implicitly organised into a unity by the life and purpose standing in their midst. They were not united to each other except in so far as each was united to Him. What was done was not the act of so many units in combination. It was the act first of Christ, and then of a living community capable by a common soul of a unitary act. These disciples, forming the first Church, were not a faggot, but a tree ; not a basket of summer fruit, but a cluster on the true vine.

Further, it is a *responsive* act, not merely reminiscent but reverberant. Its quality is fixed by the act it answers. It is a response in kind to the central, unique, eternal act which makes the Church, viz., the death of Christ, as something once done and ever doing. Its nature is not met by sitting round a table or kneeling at the altar, partaking of the elements, and calling the history before our moved minds. That might commemorate the Last Supper, but it would not re-echo, it would not show forth, the Lord's death. The true response to such an act must be another act more after its own kind. A history may be commemorated by a feast, but it is really followed up by acts done in its own nature. The Last Supper itself stood for something else ; and it is that something else which has its own note returned in the Communion rite.

The Communion, then, is more than either contemplation or commemoration. " Do this " makes it an act and not a meditation. It gives a moral value to its spiritual quality. " In remembrance of Me " seems indeed to stamp a mere commemoration-sense upon the rite—till the longer word in the .phrase shrink to its true place for us behind the mightier and the less. " In remembrance of ME." Everything about the remembrance turns on the Personality to be recalled, and the action in which that whole Personality

took complete effect. "*Accedit verbum ad elementum et fit
Sacramentum.*" But the word is really Christ, and Christ
as God's organ of grace and redemption—Christ in His
eternal redeeming act of the Cross. The precept therefore
sounds, Do this in remembrance of Him who first did this,
who gave Himself to begin with (for God asks no sacrifice
which He cannot inspire and has not outdone), and who
put His whole self into this act and gift. Do it remembering Him who is with you always; always, therefore, doing
this, giving Himself in an eternal act which utters His
whole self, ever crucified, ever broken, ever poured, ever
rising, ever gathering, by His Spirit and Kingdom, all
things into the immortal, infrangible unity of His own
infinite Person. Do this, therefore (we are carried on), in
remembrance of Him who, continuous in our repeated act,
offers Himself to both Church and world as its broken
Redeemer, ever living, ever acting Himself out, ever renewing in time the indelible nature of His eternal, crucial act,
because ever present and prolonged in this responsive act
of ours which His work stirred and inspires. If the Communion is inspired by the continuous Cross, it must be an
energy, a function, of that Cross. And as the Cross means
the Crucified, it is a function, not merely a memory, of the
Lamb slain. There is an act of the Lord Himself in our
Communion—not merely a visitation, a presence, but an
act. The Cross is not merely remembered, but re-enacted.
Not indeed in any sense in which the sacrifice is offered
afresh to God. That is one of the many ways that lead
to Rome. But the sacrifice, offered once for all, functions
afresh (if I must use a disagreeable phrase). It presents
itself afresh. It writes itself large in the history of the
Eucharist. Christ presents Himself as the crucified, in-
tuitively on Calvary, discursively in the Sacrament. But
He presents Himself in the latter less to God than to man.
Not to God, for the sacrifice once in time offered to the

Eternal was eternally offered, and once for all and for ever so far as Christ's action was concerned; nor *for* man, which is the precious finished thing in the fontal act; but *to* man, which is the nature of its repeated manifestations through the historic Church and *its* action. We must recognise more than a real presence of Christ in the Sacrament, namely, a real act. If He is there (and we believe He is there), He is not inert. He can never be inert where His Cross takes effect. *He* is in action, His death is in action, and not a mere influence from Him. He in His death is acting through His Church upon men. His determining[1] action upon God was not, and is not, through His Church; but His action on men is. And in a central spiritual act like Communion He is especially acting so. It is a great practical evangelical sermon, practical in the great sense (so desirable in more sermons) of *being* a real act, and not simply leading to acts. In Communion it is not simply that we offer ourselves through the Eternal Spirit to God in grateful response to the offering of Christ, but Christ actually and historically offers Himself as crucified to us and to the world. If the world to-day can crucify Him afresh, surely He can offer Himself afresh in the midst of it. He does so in the continuity of His body, the Church.

There is a deep distinction between this and other acts of worship like prayer or praise. In these we chiefly go to God, but in Communion God chiefly comes to us, and speaks to us and through us. In Communion there is more that is akin to preaching than there is in prayer. It is the *enacted* word of the Gospel. Christ in our act, (which is His more than ours,) offers Himself, offers His great offering, to the Church and to the world. The very commemoration of a Christ *who is our life* is worlds more than

---

[1] It will be pointed out later that in the Sacrament there is an offering of Christ to God by the Church, and so an action upon God in a guarded sense.

commemoration. It must be an act in the completed life
of such a Christ Himself. And if so, it is in some sense the
action of His death. For His own remembrance of His
death must be, to Him whose thoughts are acts, in some
sense a re-enacting of His death. And not the less so
because it may take place in our communal experience,
however inadequately conscious we may be individually of
all we do.

At any rate, to realise our Christ as also our life, in any
form of Christianity which holds to Sacraments at all, is
fatal to the bare, hard, Swiss, burgher Zwinglianism,[1] the
soul-sterilising, and Church-destroying memorialism which
starves and palters with the rite without the courage either
of taking it in earnest or of letting it go. Such paltering is
mere ritualism. It clings to a rite which has become little
more than a rite, and is slowly ceasing to be either a
pledge, a seal, or a power. It has neither the mystic depths
of Luther nor the real insight of Calvin. It is simplicity
of the wrong and thin order, like Theism, dwarfed to meet
the individual, pietist or rationalist, instead of rich to meet
a Church, or full to fit a Revelation of grace. It is salva-
tion debased to common sense, faith dropped to the bathos
of the plain man, piety desiccated and blanched by mere
polemical intelligence and attenuated by excessive protest.
And it lowers the whole pitch of piety and worship in any
community where it becomes the key-note. It prepares the
ground for the priest by stirring a need of the soul which
the priest at least recognises and attempts to fill. And so
it makes sacerdotalists by the soulless vigour and rigour of
its protest against Sacerdotalism. It is not possible for any
Church which has its experienced life in Christ crucified to
go on thus teaching the Sacrament as a mere souvenir. A
mere commemorative Sacrament is but the relic of a dead
Christ, and the badge of a dying Church.

The tendency to make little of this act is one which

[1] More Zwinglian than Zwingli.

exists even among many whose piety is unquestioned, but
it is usually associated with but slight regard for the Christian life as life in a Church. Some, who are drawn to
Christianity chiefly by its ethical and philanthropic side,
tend to reduce the practical act of Communion to some-
what low dimensions in order to enhance the superior
sacrament of Christian conduct, and to express its independence of specific forms of worship. But this tendency is
after all only one aspect of the alarming baldness and
poverty which have overspread much of our services, taking
the rapt soul out of our prayers, and the warm worship out
of our praise. And it is in great measure the cause of this
declension. It is because our associations with Communion
are neither solemn nor rich enough that our other worship has been so often flat and poor, our services casual,
familiar, or humdrum. And in the efforts we do make to
purify and enhance Communion we have sometimes gone
the wrong way to work. We have tried to secure purity
by testing the communicants, and the purity we get is
neither complete, nor is it imposing. We have sifted the
participants instead of subliming the rite and Presence.
Give it its true value, its most solemn interpretation, hedge
it with no fictitious rigour of precaution, but transfigure it
with a real solemnity of meaning, and it will become a self-
acting test. It will exert its native affinities, and do its
own spiritual selection. And its own severe glory will warn
off the unconsecrate in heart and soul, as from the death-
dawn in the face of Christ the soldiers fell back who would
have lifted up on Him unholy hands.

Again, something like a true Sacramentalism as distinct
from a pious reminiscence might help to cure that Sentimentalism which is so ineffective in the humaner develop-
ments of Christianity. The worst weakness of Liberal
Christianity is not that it is negative or destructive, for
it is neither ; nor that it is untrue, for it exists by the Spirit

to release the truth and undo the falsehoods of the past. It is the instinct of self-preservation in Christianity, and the habit of self-examination, which is a grace of the Spirit. But one of its great weaknesses is that it is, in so many of its more popular advocates, sentimental, feminine, and subjective. A more masculine and commanding faith would follow an increased emphasis upon the objective side of the Sacrament. *For mere commemoration must always be subjective and individualist in the main.* The reminiscence by the worshipping subject will always be more prominent than the object itself, which is not real because not present but only fetched from the ghostly past by the affection of the hour. In a true Sacrament we have an act rather than a sentiment, and an objective presence more real than any subjective state of ours.

It may be objected that what closes the door to Sentimentalism opens it to Sacerdotalism. To which the reply is, only if a magical instead of an ethically-spiritual transaction is believed to take place ; only if we lose the evangelical view of the Cross as the active ethical centre, and Redemption as the permanent ethical principle of the race, and its moral soul. The safeguard against priestism is not the attenuation of the Sacraments but their true interpretation. Our error often is to starve the idea till it lose its strange power over a whole side of the human soul, and so we drive to the priest all who need food for the spiritual imagination and are fascinated by the saddest solemnities, the most hushed pieties, and the darkest beauties of the cross and its unearthly strain.

To venture a little way into explanation, we have three pairs of terms :—

(1) The body of Christ, and the material world.

(2) The Act of Christ in His death, and the act of the human will in Christian devotion.

(3) The person of Christ, and the person of the Christian.

The truth in the Sacrament consists in the true relation among these terms.

(1) Taking the first pair, the body of Christ and the material world (bread and wine). It is here that the magical theory chiefly operates. So long as men attempt to set up in the Sacrament a real relation between this pair of terms it must issue in magic with the priest for the wizard. *Hoc est corpus*, becomes hocus-pocus. What we must say is, that with our possible knowledge we can set up no relation between these terms. About the body of Christ in this sense we know nothing. A local and spacial heaven is a representation now valuable chiefly for pædagogic purposes. That on the one hand. And on the other, we know too little about the ultimate constitution of matter. We have no knowledge which will enable us to bring a heavenly body of Christ and the material world into valid relation, or to give Transubstantiation any meaning for thought. Calvin even, who was the truest of all the Reformers on the Sacrament, seems, in his views, to have suffered much from the local and material theories of his time about the future state and the world unseen. Even he took the body of Christ and its ubiquity too literally. And it was largely due to the error, popular then as now, which understands by Spirit only highly rarefied substance and by a Spirit a ghost.

(2) Taking the second pair of terms—the sacrificial act of Christ and the sacrificial act of Christian men. The Catholic theory here is that the human act in the Sacrament (the priest's act in the Mass), is a *duplicate* of Christ's act upon the cross; especially in this, that it is a sacrifice offered to God by man rather than a sacrifice proffered to man by God. I do not say that the Catholics would admit the statement baldly made, but I mean that their doctrine amounts to this in effect (particularly with their view of the Church as Christ Himself in a permanent incarna-

tion). And I refer to the distinction between an act which simply repeats another, and one which is a constituent part and organic factor of that other, extending and actualising it. In Catholicism the two terms of the relation have an excessive and fatal independence of each other. The Mass repeats the Cross. The act of the priest has a direct action and effect in the invisible world (as when Masses release a soul from purgatory)—a directness at least so great as to compromise the mediation of Christ and aggrandise the officiating priesthood. The Protestant theory on the contrary relates the two terms in no such parallel and irreverent way. It relates them as the body is related to its members, not by way of repetition but by way of functional contribution. The human act is to Christ's act as a living cell is to the living organism. Our act of sacrifice is a vital factor, infinitesimal in its own value, but infinite in its worth as organised into that eternal life of sacrifice which is the redemptive spirit of the world. And our devotion, whether in rite or conduct, is an ethical thing, a part of our moral and spiritual constitution in Jesus Christ. It is through Christ as our Mediator—as Mediator of Universal Humanity, not as any mere individual intermediary—it is through Him (meaning *in* Him), and not directly as individuals, that we pray and act into the unseen. It is through Him that our human cross with its devotions and renunciations has any action upon the world of spirit. It is the completeness of His Sacrifice that at once requires ours and gives ours validity. He only is the one priest, and it is the Son of Man's sole and sufficient priesthood that requires that we should be priests to be men. Just as because He lives we live also. It is one of the functions which go to constitute His life, and are by that life made possible. If Christ be not our life, but only our teacher, our example, or even our ideal, then it is but metaphorical to hold speech of this sort.

In no real sense is our act His act. Nor do we in our
cross reproduce an Eternal Cross from within, but only
imitate from without and afar a historic martyr. Our
act in that case has but an external and accidental connexion with His—a historic connexion, if you will, but
not any connexion organic, or, in a real sense, spiritual.
In the *real* sense it is not spiritual. It may proceed from
a spiritual temper and affection, but not consciously from
the ultimate spiritual ground. For religion it may be
spiritual, for the deeper considerations of positive faith it is
not. But if Christ be our life then our act is His act, our
life as practical is one with His as practical. And we are
not only at one, but we are one. We not only commemorate His act, or even imitate it, but we *do His works*. And
so strong an expression is only justifiable on the ground
that it is not we who live, but Christ that liveth in us.

(3) This brings us to the third pair of terms, and to the
relation between the person of Christ and the person of the
Christian.

It is in this region that the real union and transubstantiation takes place. The body of Christ really and
finally means the person of Christ. Bread and wine are
symbols of the flesh and blood in which matter is raised to
an organism. But flesh and blood are themselves but
symbols of an organism higher still, the organised personality, the Spirit. "They two shall be one flesh," means
one spiritual personality, slaying the spirit of individualism.
And we are reminded of the saying that in Christ is neither
male nor female, because He is both, because He is the
universal personality in whom all individuals are saved,
and gain their individuality by losing their individualism.
That is to say, in simpler words but more enigmatical
phrase, they gain their souls by losing them.

The essence of the Christian life is personal union with
the person of Christ. There are Christians who suspect

such phrases as these of mysticism, who dislike mysticism, and who accordingly explain the phrases away, or simply ignore them. But they will not be ignored. And fortunately the sole alternative is not mysticism. If common sense, with its rough methods, thinks not of union but merely of attachment, the mystic is apt to err in the other direction, and think not really of union but of fusion, which is a very different and more dangerous thing. The mystic is often a pantheist without knowing it; he loses his self without finding it, and merges in the general soul. His piety loses both measure, modesty, and virility. The word of the cross, however, is Reconciliation, and its end is a Union which subsists upon the ethical conditions of fixed personality, and upon an intimacy of communion and being far profounder than is possible by any crude ideas of mystic fusion or personal erasure. " Christ liveth in me " may be the word of the Christian mystic. But the word of the Christian saint and apostle of Reconciliation is, *" I live, yet not I, but Christ liveth in me."* The life of Christ is the ground of the Christian life, not its substitute, nor its mere material. This phrase of Paul's is the key-note of the Christian's experience. The real objective ground in true Christian life is the person of Christ. This is the Real Presence, substantial but not corporal, spiritual but truly objective. We have a communion not of act only, nor of work, but of life and being; and Redemption and Faith are, so to speak, but the two poles in one completed spiritual sphere.

The person of Christ is our true objective. But the key to the person of Christ is in the cross; for the cross is the principle of God's Revelation, no less than of our Redemption. The cross is the bond of all bonds, the unity of all unities. It consummates the internal unity of God. It consummates the internal unity of man. It consummates the unity between God and man. The grand bond be-

tween person and person, heart and heart, is the cross with
its renunciation, its sorrow, its holy, atoning power. Just
as between husband and wife, for example, no common joy
deepens the bond like the loss of a child, or the danger of
losing each other. The key to the depths of our personal
union with Christ is the cross and the fellowship of His
death.

Hence the rite of the cross has a special and unique
significance in Christian culture, in the working out of the
union set before us.

But is there a special presence of Christ in this rite?
The expression, special presence, like special providence, is
if not self-contradictory, at least unhappy. It is always
the same unchanging Christ, who never leaves us nor forsakes us. It is the same Christ in our prostrate worship as
in the minor awe of our reflection, and in the sobriety of
our walk and conversation. But to the question so put it
is more true to answer yes than no. It seems at least a
different presence—the same Christ in a different presence.
Perhaps a better expression would be the more immediate
presence. But is Christ more immediately present in this
rite than in the depths of our solitary communion with
Him? Yes, the whole, the divinest Christ, is—the Redeemer of the world and not of our single soul. The
speciality of the presence in the Sacrament is the *community* of so near a *Redeemer*. It is the universal Saviour,
the common Christ, that we worshipfully realise, not the
individual's. And Communion differs from other acts of
common worship in this—in the solemn immediacy of His
common presence as Redeemer. He is as immediate as in
private worship, and as universal as in public. Hence the
Sacrament is the blessed mean and meeting-point of public
and private prayer. In private worship we are apt to be
self-engrossed. In public we are too dependent on the leader
of the devotion, or the preacher who strives to kindle the

common flame. In the Communion (especially if it be to any extent liturgical), the leader sinks away, becomes but the voice, becomes the echo of a voice, whose echoes have been multiplied in every age, the channel (although the living channel) of the voice of Jesus walking in calm light upon the world's wild waves, "Come unto Me." As a community we are then in the immediate presence of the Universal Redeemer, the real presence, as Calvin says, and yet not the local presence, as Zwingle truly against Luther declares. And the elements, while they are *signa*, are no more *nuda signa*, or bare suggestions, but *signa mystica*, not indeed changed into what they signify, but lost and irradiated in a halo or corona of spirit, visible only to the eye assisted by faith.

But if this be so, then the true doctrine of the real and immediate presence of Christ in the Sacrament, so far from opening the door to priestism, is of all doctrines that which makes priestism impossible. For it is there we realise most the immediate universality of Christ in the Church as Saviour. We have each our equal ground in His sufficiency, and because He is complete we are, each one of us, alike indispensable. We realise there especially the unity of men in His Redemption, His immediacy to each soul in the common presence, and the consequent impossibility of a privileged Sacerdotal caste with a magical prerogative or a historic commission to mediate between Him and us.

Finally, we shall thus preserve the real and powerful objectivity which is the truth whereon priestly superstition builds; we shall give our doctrine that air of positive actuality which meets a realistic age; and, on the other hand, we shall exert more influence than we have done upon the beautiful night-side of the spirit, we shall feed the starving spiritual imagination, and stir the trembling praise from the shadiest coverts of the wounded soul.

P. T. FORSYTH.

(To be continued.)

SACRAMENTALISM THE TRUE REMEDY FOR SACERDOTALISM.

II.

We are robbed of some due sense of the true place of Communion by our misuse of the word symbol. The elements alone are not symbolical. Symbol, at its best, is something that not only *reminds* us of reality in the significate, but by its living nature *passes us on* to the reality. Then Communion is organic, and not arbitrary—not a mere matter of association. It' is not through the mere elements that we touch the reality, else Rome were right, and we are lost in all the metaphysics in which transubstantiation has smothered faith. The elements are but the material which the true symbol employs in passing us on to . the reality. That reality is in the region where all reality must accrue at last, and be found for ever at home with itself, in the region of will and of action. It is, of course, the person and work of Christ. Now the elements are not so symbolical of this as is the action performed on them. It is the breaking, the pouring out, the partaking that are the true symbols. That is to say, the true symbol is not an element, but an act. It is only thus that it can be a symbol of the great act which is its reality, the act of the cross. It was so at the Last Supper. It is so in our Sacrament. The symbolism is in the Church's act. It is therefore a symbol which itself belongs to reality. It has the reality of will—of our will, and of Christ's dying will acting through ours.

There is no fear of any superstition in emphasizing *this* real presence, so long as we urge that it is a reality of present act and will, and not of mere substance. We renew our first decisive dedication of ourselves to Christ, and Christ renews His first decisive offering of Himself for us.

It is a real renewal of the devoted act ; and it is equally real on both sides.

Much of the strife that has arisen about the Last Supper might have been avoided, and much may be laid, by a true grasp of the principle by which the Old Testament explains the New. The Old Testament explains the New as the New Testament lights up the Old. The Old Testament interprets the New ; the New Testament reveals the Old. We cannot understand the Old Testament without the New, and we cannot account for the New Testament without the Old.

The best clue to this act of Christ is in the Old Testament; and it is in that part of the Old Testament which was most in Christ's own thoughts, and is therefore most fertile for understanding Him and His work. It is not in the law, where it has been sought to excess and to strife, but in the prophets. The New Testament men altogether were not priestly, but prophetic in their strain.

The key to Christ's intent will therefore be best found perhaps in the method, used so often by the prophets, of symbolical action. The overladen thought passes beyond the power of words (as thought inspired by love and passion must at its height always do), and is driven into the symbolism of an act. It craves an enacted instead of a spoken symbol ; a parable in startling deed instead of stale word. Love surcharged passes through the broken alabaster into silent sacrifice for its full vent ; and inspiration at its height forsakes the word and takes up the work. Signs become more eloquent than speech. To threaten calamity and captivity, with a force for which words had failed, Jeremiah lays a yoke on his shoulder, and Isaiah goes barefoot. To express victory another puts on horns, the symbol of power. To represent the rending of the kingdom, Ahijah rends his garment and gives ten pieces to Jeroboam. In like fashion the events and calamities of the prophet's domestic life cease to be private, and become prophetic symbols of public affairs, as with Hosea and Isaiah. The cases are numerous enough, and not unfamiliar. We only move along the same path when Mary meets us with her costly spikenard and her tears. We go farther, and find the Saviour Himself kneeling in masterful humility to wash the disciples' feet. And at the end we look into the upper room and behold the Last Supper, the incipient Passion, and the symbolic act in which the burthen of His gathering agony found relief. This was, as has been said, " Christ's last parable." It was a parable translated now from word to deed—a twin parable (as the Lord was used to group His parables in pairs) by the action with the bread and with the wine. The word that constitutes the Church was a deed. *Im Anfang war die That.* The divine Teacher had done His work, and was rising into the divine Doer, the Redeemer. The lesson, taught but unlearnt, must now be conveyed by an action which will not fail. The great act of the Cross was impending, of which only another act, and not a word, could be the symbol. The central point, therefore, in the Last Supper is not the symbolism of the elements, but the symbolism of the action. It is on this line only, perhaps, that we can hope for a happy issue from the vast controversies that have gathered here.

The symbolism does not lie in the elements, but in the act. That is the exact point. To remember Him was to " do this," to " take and eat." The stress of the situation falls not on " body," but on " broken "; not on " blood," but on " shed." What was symbolised on the occasion was not a mere manifestation on the cross, but a decisive act there ; something not only exhibited, but done. Revelation is Redemption. Wherever our thought wanders from this aspect of the cross, and sees in it only a declaration, or an epiphany, of the love of God, the Sacrament shares in the loss of tone. A theology of mere revelation produces a

Church of mere sympathies. It fails in faith, sanctity and power. And amid a disillusioned world the Church sinks, sweetly vapid and witlessly content, to its amiable, ignoble end.

(1) We note first, then, that it was an action that was to be symbolised. It was the work done for us by Christ—our Redemption. The eternal Christ, who is an everlasting Now, anticipates in the Supper His finished work, and in symbol says "it is done." The value of our Lord's actual flesh and blood was little before God. It was in no symbols of these that the sanctity lay. It is only metaphysical theories that have made them of such account. *The* precious and sacred thing was His holy, God-beloved will and its complete obedience of faith. There is the nerve of personality, there is the seat of sanctity. There the great, eternal, final Redemption transpired. The value of the cross lies in its value as an act of Christ's soul and will. That act was the thing to be symbolised.

(2) It was, therefore, an act which symbolised it : it was not the elements. An act is a spiritual thing. Its truest symbol is another act. The elements are no more than materials to enable the symbolic act to be done, as the body itself is but a finer material in the service of that act. When shall we take it fully home that as the Incarnation was not a physical act in the first place, so neither was the Atonement? The accent falls neither on the physical entrance of Christ into the world in the one case, nor on His physical sufferings of exit in the other. The secret of the Incarnation lies in the personality of Christ, whose centre is the holy Will.

And we may illustrate thus. A spoken word is the symbol of a thought: the visible letters only enable us to convey the symbol. They are not the symbol itself. What the letters are to the word, that the bread and wine are to the Sacrament, στοιχεῖα, *litteræ, elementa*. What the word is to the thought, that is the Sacrament to the cross. Only that the Sacrament, as it symbolises not a truth or thought, but an act, is an *acted* word, a deed, the community's response in kind to the act that made a community of it; and being an act, it has a reality in it, symbol though it be, which no material elements could have.

We repeat the word often; the thought is there once for all. The music is performed often: the composer's work stands there as a spiritual totality of achievement, render it as often as you will. We repeat often the symbolic act, but the work of Christ which is rendered in it is done once and for ever. That work, in a true (if guarded) sense, repeats itself in us when we obey in the memorial act. It is misleading to speak of the action in the Sacrament as merely symbolical, and not reiterant at all. It is not symbolical in any sense that would impair its relative reality. As the Romanists, with their false start from the elements, are forced to place under them the Lord's real body, so, starting from the true base of the action, we must own in it the real acting of the ever present Christ, the real operation of His work and cross, the real self-utterance of His undying death. It was the same will, in the same effort, that both died and enacted at the Supper the symbol of His death. And it is the same death which acted backwards, if we may so say, in the *institution* of the Sacrament, and which acts onwards in our *observance* of it. The Last Supper and Gethsemane forefelt and foredid the cross; rehearsed it, if such a word may fitly be applied to anything so absolutely real and so little dramatic in each case. Neither was a mere rehearsal, any more than our observance is, a mere repetition or commemoration. It is the same act and will uttering its fundamental reality in both, in its preludes as in our aftersong.

(3) We have, therefore, really a symbol behind a symbol. The broken bread stands for the broken body; the broken

body stands for the broken, bowed, but invincible will. The ultimate reality is the will's act. The great symbolism and sanctity, therefore, must be sought in the *breaking* of the bread and the *breaking* of the body, and the *partaking*, not in the bread or the body as elements *per se*. The true vehicle and symbol of an act is not an element, but a living body capable of acting. A substance might symbolise a substance, as bread the body. But only an act can symbolise an act, the material act the spiritual. That which is born of the flesh is flesh, but that which is born of the Spirit is Spirit. But every act is a spiritual thing. It is an act that the symbolism ends in, and therefore action is the region it all moves in. The acted symbol, especially at the first supper, is thus more than a symbol. It is part of the reality symbolised. It is the utterance of the same act of will. It was the same will that broke the bread and bowed on the cross. And it did both in the strain and exercise of the one spiritual act that redeemed, the *actus purus* extended through Christ's total personality as its characteristic energy. The symbolism of the occasion, I repeat, lies in the action, not in the elements; and the real presence is the present action of the Saviour's will, not of His substance. It is there not for contemplation or adoration so much as for communion. We all hold to Christ's real presence in Communion; but if it is not in the substance, it must be in the act. The real presence of Christ is not in the elements nor in the air, but it is His act within our faithful act. Christianity means nothing if spirit cannot thus interpenetrate spirit, and act act. It is not on the altar He dwells, but in the common will surrendered and united to Him. It is not in the temple space, but in the community of the obedient Church. This points to a Sacramentalism which is much other than commemoration, and yet is the deathblow to Sacerdotalism. It ends the worship of the elements, and the monopoly by the priest of the consecrating function of the community. It is the faith of the present community that completes the act. The essence of the Sacrament is the common act of the common indwelling Lord, and the symbolic act ceases to be symbolic merely. It is profoundly real, and therefore alone profoundly religious. Our worship is no more subjective and sentimental, as commemoration must become. It is positive and objective. It is the act of God in its return to God; the Holy Spirit in sublime death returning to God that gave it. Every act of a revealing God is reflex, and is incomplete till it return in congenial response. The finished work includes a Church and a Church's acts.

The action is real on both sides. It is a real assignment of ourselves to Christ crucified; and it is a real assignment by Himself of Christ crucified for us, as I shall shortly show. I quite accept the old illustration given by Dr. Dale, and the validity of its distinction between a surrender of the keys by the governor of a besieged town, and a ceremony in which the forces of the besieger present him with keys emblematic of those he has won or is to win. It is a case of real offering and surrender in the Sacrament, both on our side and on Christ's. It is not dramatic, not ceremonial, not commemorative alone. As Christ *was* God's act of grace, and did not merely announce it, so our central worship is a real offering in return, and not the mere expression of surrender effected somewhere and sometime else. We offer ourselves anew. We utter in a solemn detail and special function the compendious act of consecration, which is the standing and decisive relation of our soul to God in Christ.

(4) But we do more. Such a view is still too subjective; it tends to be too introspective, and ends by being too sentimental. We make a more objective offering. Something in our hands we bring—something not ourselves which makes our righteousness. We bring Christ, and

offer Him far more truly than is done in the Mass. The
great hold and power of that rite is due to the objectivity
of the offering. This overrules for many a soul its falsity
in that which is offered. Well, we do not offer His body
and blood, but we do offer Himself and His act of death.
We make His *soul* an offering for sin.

Men once offered Him up on the altar of their rage and
hate. Man will go on now to offer Him for ever on the
altar of his repentance, gratitude, and adoration. We have
nothing else to give, and worship is giving. We can but
bring to God what He has provided. What is the value
of our sin-stained thanks in themselves to Him? What is
the worth of our mere emotions, our faltering resolves?
The broken, contrite mood is not necessarily the contrite
heart which has broken with self and sin. What at least
is the value of these things as a return for all that is meant
by grace, forgiveness, redemption? We are not worthy
even to thank Him but in a worthiness He Himself gives.
That worth is Christ in us, in our praises, thanksgivings,
Eucharists. It is only Christ in our praises and prayers
that makes them worship. This is a truth which may
seem to æsthetic, literary, or (most odious of all) stagey
piety both narrow and inhumane. And, indeed, to a religion which is in the first place humanist and only sympa-
thetic it must so seem. The sorrow of Christ is the agony
of a strait gate. But it is mankind's only avenue to the
Kingdom of Heaven ; and it is this kingdom, and not
Humanity, that is the ideal and principle of Christian faith.
And the kingdom of God draws its value from Christ and
Christ's death. The prophet was hallowed by the king-
dom, but the kingdom is hallowed by the Christ. It is
He in us who consecrates any feelings or deeds of ours to
God. We have nothing to offer God but Christ and His
Cross. It is not our warmth of feeling towards God that
makes it welcome to Him, nor our obedience of act, nor our

sincerity of intention. This *is* the work of God to believe
in Jesus Christ. It is our warmth, strength, or reality of
faith that wings our worship. And faith makes us feel
that no penitence, praise, prayer, or sincerity of ours is
worth anything to God as worship except in the midst of
them there is the Sacrifice of Christ once offered in
time, and in the world of spirit continually being offered,
especially in the life of souls dead in His death. In all our
worship we are but giving Christ back to God. We are
making His soul an offering who first offered Himself as
God's offering. We are not simply remembering Him, but
renewing in our spiritual experience that perpetual experience of His in which by faith we share. Our union with
Him aspires to share His spiritual experience to-day, an
experience in which the cross of Calvary is surely something much more integral and potent than a reminiscence;
while its expression by us is for Him who acts through us
surely far more than a memorial. His intercession, as the
prolongation of His redeeming act, is surely more than that
He—

> " Still remembers in the skies
> His tears, His agonies, and cries."

All this is especially so in partaking of the Sacrament of
His death. We are made priests unto God. We take
Christ's offered soul in our hands, as it were, and offer Him
to God, in no material fashion but in our redeemed experience as wills united with Him. All communicants
have not come to realise this height of the matter as
yet. They have stages to run, and initiations to undergo.
But such is the goal and idea of the Church's Communion.
We make His offered soul our soul's offering. We hallow
into worship all our subjective experience by His objective
work and its real presence. He not only stirs our emotions
by His memory, but being in us, mingled with our experi-

ence, He consecrates them and carries them to God. He makes worship of them by creating them, and by incorporating our act with its parent act, with the sole, sufficient, and all-hallowing act of worship ever done to God, namely, His own soul's obedience, agony, victory, and praise. No religious excitement or energy is worship till sanctified thus, either within our knowledge or beyond.

(5) But I would go farther still, and say that in the Sacrament we have a real offering from Christ's part also. We can never, never hold against the sacerdotal churches till we are sacramentarian enough in our worship to go beyond them in the reality of the offering by exceeding them in its truth. We must offer, as I have said, not ourselves only to Christ, but Christ Himself to God. But also, going farther, we must furnish opportunity for Christ's renewed offering of Himself through us to the world. We have to do more than announce His gospel. We must transmit it. We have not only to preach Him, but give Him effect. We cannot redeem men to God, but we can do much to reconcile. That is a great sacramental function. It is Christ acting through His Church on the world. And with most Christians and many churches life is so little sacramental in its tone and reconciling in its effects because we are so far below the sacramental in our central worship. Our weakness before Rome and all that is Romeward lies in the poverty and subjectivity of our sacramental faith. Our churches are not in earnest with a sacramental view of life because we are nervous about a sacramental view of worship. We are more afraid of the priest than sure of the Presence. Mere protest is conducting us through Zwinglian attenuation to Socinian negation. We do not act in worship or life as if we were men in whom Christ crucified is offering Himself to the world, through the Church as its hope. We turn often from the sacraments with an impatience so rugged that it is more self-willed than honest,

and we say we will not observe them but live them. And
certainly we succeed so far as that our living of them is
without observation.

The Communion is an act of the Church moved by
Christ in its midst. But if He is present in the act to
which He inspires His Church, then He is acting by His
Church, He is doing something. And on such an occasion
that something can only be in some real sense the act of
the Cross. The Cross is the central energy of His spiritual
world, the focus of all the influences that constitute the
kingdom of God. It is the real point of departure for the
Holy Spirit, even if the resurrection was the point of
emergence, and Pentecost the point of attachment for the
Church. In such an act of the Church, therefore, Christ
is in a real sense offering Himself. He is at least offering Himself continuously *to* the world as the Crucified,
who was once, but for ever, offered *for* it. The Sacrament is always some real function of His Sacrifice—that
is, of Himself in sacrifice, and not simply of us in response. It is a great act of preaching by the Church,
which is the hierophant of an undying inspiration. It is
practical preaching in the great sense of the term—which
(as I have said) is not, in the day's phrase, preaching " conduct," but preaching by a great act, by a word which is
really a deed, as the gospel word in its essence is. We do
not repeat His Sacrifice as the Mass professes to do, but
we do re-echo it in the only way an act can be re-echoed
—by another act in which the initial act returns upon
itself in kind as a real act of spiritual will, and not of
institutional ritual. The priest *offers* a real sacrifice in
each Mass. We in each Communion but *proffer* the real
sacrifice offered once and always by Christ alone. But it
is His offering all the same that is the active and efficient
element in our proffering. His action is our real presence
and power. We are not mere participants but factors in

the mighty act. It is by an act which is ours, but also and still more, Christ's own act in us. It is the living Christ re-asserting by act, through the Church which His death made, that one unique, infinite, sufficient death, never to be repeated even by Him, yet never to cease acting and reproducing itself in our will and deed. His death is, in our act as a Church, not simply recalled, not simply related, not simply witnessed to by us, as a report of old, forgotten, far-off things. To show forth the Lord's death, is, in a sense we are too timid about grasping, to re-enact it, to let it re-present itself in us as real action within real action, a real presence in real effect where the last reality lies—in the spiritual will. It is an act and energy of Christ Himself if He be His Church's life, if the outgoing focus of His life in the redeemed community be the act of redemption, and if the ingathered focus of our worship be the rite in which we act purely and only as souls redeemed. It is a function of Christ's ever vibrating act of present, undying death ever offered through the Church in the heart's region of spiritual reality to the soul, to the world, and to God.

The acting subject in the Sacrament, then, is first, Christ, and, second, the Church. "It is God that baptizes us," says the Apology for the Augsburg Confession, "and the minister only in His name." And the like applies to the real agent in the other Sacrament. But the Church acts as a community of individual believers. And on the part of each soul there is action which, symbolic as it is, is not prophetic or predicative, as the act of the community is, but appropriative. In the act of consuming the elements there is a symbol of that union between the person of Christ and the believer which is the soul of Christian faith. And it is a symbol which is not mere symbol, but such a function of loving union that in the act of commerce the reality is consummated and deepened. Here, again, it is

not so much the elements that are symbolic but the act.
It is not the substances that meet—the spiritual substance
of Christ under the elements and the spiritual substance of
our soul. Such an idea is really materialist, however re-
fined. It turns sacramental grace into something that can
be infused in a sense too literal for spiritual safety. It
opens the way to believe in an infusion of grace which
incorporates it with our nature in a sub-conscious region
independent of any intelligent spiritual activity of ours.
The mysticism then becomes magic. We are transmuted
without being converted, consecrated without being sanctified. It is not thus that grace works. It is not the Saviour's
corporeity that is conveyed, however glorified. It is His
Person and work acting from the eternal world on our
person in its responsive work and receptive energy. Spirit
with spirit meets, life with life. His flesh means His personality, His blood its distinctive native energy, namely,
His redeeming work. It is on these we feed. His spirit
and energy pass into ours in conscious communion. What
meets is here again two wills in an act, two personalities
in blended function. We may call this union mystic if we
will, but it has none of the dangers of a mysticism con-
ceived as the blending of two substances, however ethereal.
It is intelligent, interpersonal, not fusion but interpenetra-
tion, the union of two moral beings in an act which is none
the less a moral act that it transcends the limits of such a
term. It is spiritual in the sense in which only beings of
a moral nature destined for love and trust can be spiritual.
It has the spirituality possible only to living persons. We
appropriate Christ in the Sacrament, therefore, in no other
way or measure than as we appropriate the gospel, the
work of Christ for the conscience and on the conscience.
The Sacrament of the word is the key to each Christian
Sacrament. They exist for the sake of the word of the
gospel. They have value according to the extent to which

they are charged with that and give it effect. And what the Lord's Supper conveys is not only the word made flesh, but still more made sin for us, the word as a living, acting, redeeming personality, in contact with our faith. What it effects is this union with the like personality in those who partake, who are forgiven, and who become the righteousness of God in Him.

It is the gospel which interprets the Sacraments, not the Sacraments the gospel. That is the grand principle of a a Protestant sacramentarianism. The Sacraments depend on our idea of Redemption, on our kind of faith.

If we thus fix our symbolism on the proper point, and find it in the act rather than the elements, we gain two things. We transcend the jejune idea of a mere commemoration, upon which no Church can live, however a school, sect, or society may perpetuate it. And we escape from the evil sacramentalism which historically goes hand in hand with priestly prerogative, and which philosophically materialises heavenly things by spiritual ideas really drawn from the qualities of substance. It is impossible in course of time to escape the dangers of either extreme. Commemoration dries into lean Socinianism and a piety of parched commonsense. And the veiled materialism of the Mass appears in the general soul as a paganism and superstition which are a correct translation of the false sense underlying all.

A profound sacramentalism is the only exit from a false sacerdotalism.

And the writer cannot veil his conviction that much objection to it is more polemical than positive, more protesting than informed, and that it proceeds, in many pious cases, not from spiritual freedom, volume, or vitality, but from the autodidact's lack of spiritual depth, seriousness, and sequacity of thought.

P. T. Forsyth.

www.ingramcontent.com/pod-product-compliance
Lightning Source LLC
Chambersburg PA
CBHW022033190326
41519CB00010B/1703